THE LIVING BIOLOGY ANTHOLOGY

A Confluencist Testament In The Living Science of Creation

VOLUME I

LIFE, LIGHT, AND RENEWAL

Written & Curated by George Powell
Illustrations by Nick Jones
George's Media LLC

THE LIVING BIOLOGY ANTHOLOGY — VOLUME I

Written & Curated by George Powell

Illustrations by Nick Jones

Design by George's Media LLC

ISBN: 979-8-9939108-2-6

Printed in the United States of America.

EPIGRAPH

"For in Him we live and move and have our being."

— Acts 17:28

"Creation is alive with meaning for the one who has learned to look."

— Confluencist Reflection

DEDICATION

For all who wonder.

For all who question.

For all who see science not as a contradiction, but as another language of the Holy.

This book is for the curious, the quiet, the bruised, and the becoming.

May you find yourself in the living world and find the living world in yourself.

ACKNOWLEDGMENTS

First, to my mother, my first teacher in life, whose care, patience, and example formed the ground beneath my curiosity.

Before there were lessons or laboratories, there was her steady presence, teaching me how to observe, to listen, and to tend what is growing.

To my sister Angela, my first science teacher, whose love of discovery shaped my earliest sense of wonder and planted the seeds for this work.

To my colleagues and students — the members of the Salesian Boys & Girls Club, the students of The Math Science Investigators and Huguenot High School, and especially my Science Department family — thank you for the years of curiosity, laughter, questions, and shared devotion to helping young minds awaken.

And to Nick Jones, whose luminous illustrations breathe spirit into science and help creation speak.

Finally, to all teachers, parents, healers, and seekers who believe that knowledge is an act of love and learning is a form of re-creation.

ABOUT THE LIVING BIOLOGY ANTHOLOGY

Living Biology is a Confluencist exploration of life — a meeting place of biology, spirituality, ecology, and story.

Each chapter carries:

- A Witness from the natural world
- A Reflection interpreting its spiritual rhythm
- A Sacred Margin or Sacred Stream guiding practice
- A Charge for daily living
- An illustration expressing the chapter's essence

This anthology continues the work of The Living Testament Anthology, bringing the same sacred curiosity into the study of life itself.

TABLE OF CONTENTS

Chapter 1

Sacred Scientific Method

Experiments of Faith & Observation

Invocation

O Eternal Teacher, open my eyes to see that all inquiry begins with wonder, that all knowledge rests on humility, and that every question is a pathway back to You.

Witness — What Is the Sacred Scientific Method?

Before microscopes and telescopes, before laboratories and equations, humanity practiced inquiry in dialogue with the Divine. The Sacred Scientific Method is older than science and deeper than formula. It is the discipline of:

1. **Observation** – beholding the signs in creation.

2. **Testing** – placing our questions before reality itself.

3. **Reflection** – seeking meaning in the results.

4. **Humility** – acknowledging mystery, not conquering it.

This is not a method to control the world but to encounter God within it.

Witness — Scriptural Examples of Sacred Inquiry

» Daniel's Trial (Daniel 1:12–15)

» Gideon's Fleece (Judges 6:36–40)

» Wisdom of Solomon (Proverbs 6:6; Ecclesiastes 1:5–7)

» The Voice to Job (Job 38–39)

» The Qur'an's Call (Qur'an 3:190–191; 88:17)

» Muhammad's Wisdom (Hadith, Tirmidhi)

Witness — Applied Sacred Method (Lab Reports)

Lab Report 1

Title: The effect of a plant-based diet on physical vigor among royal trainees (Daniel 1:8–16)

Independent Variable (IV): Diet type (vegetables + water vs. royal rations of meat/wine)

Dependent Variable (DV): Physical appearance/vigor after 10 days

Hypothesis (If–Then): If selected trainees consume only vegetables and water for 10 days, then their physical appearance will be as good as—or better than—peers consuming royal rations.

Background/Rationale: Daniel seeks to remain faithful to covenantal food practices. The steward agrees to a time-bound test with visible criteria.

Methods:

- **Participants:** Daniel, Hananiah, Mishael, Azariah (treatment); palace trainees (comparison)
- **Design:** Short-term intervention; non-randomized
- **Duration:** 10 days
- **Materials:** Vegetables, pulses, water; royal rations
- **Procedure:** Treatment group consumes vegetables + water; comparison continues the usual diet
- **Measurement:** Overseer's inspection (appearance, vigor)

Observations/Results: After 10 days, the treatment group "appeared better and fatter in flesh."

Analysis: Supports hypothesis; limitations include non-random assignment and subjective outcomes.

Conclusion: Plant-based regimen associated with equal or superior vigor.

Limitations & Next Steps: Consider objective measures and randomized design.

Sacred Reflection: Faithfulness operates like a hidden variable; devotion interacts with practice.

Lab Card 2

Title: The effect of fleece condition requests on sign confirmation (Judges 6:36–40)

IV: Requested dew pattern

DV: Observed dew distribution

Hypothesis: Dew pattern will match request if sign is divine.

Method: Two-night reversed-conditions test

Results: Both patterns occur as asked

Conclusion: Sign confirmed

Notes: Functions as two-condition verification

Lab Card 3

Title: The effect of collection timing on manna spoilage (Exodus 16)

IV: Day/amount of manna collected

DV: Spoilage rate

Hypothesis: Hoarded manna spoils; sixth-day manna endures

Method: Community instruction, next-day observations

Results: Ordinary-day manna spoils; double-portion persists

Conclusion: Timing modulates perishability

Notes: Sabbath as rhythm variable

Reflection

Modern science is but the younger child of this older way. The Sacred Scientific Method does not strip away wonder — it sanctifies it. To ask questions of creation is to converse with the Creator. To test and to reflect is to worship.

Sacred Margin

Hold one question today. Ask it not with arrogance but with reverence. Look to creation, to scripture, to your own life for signs. Let the answer humble you, even if it is silence.

Charge

Remember:

Every experiment, every observation, every reflection is holy when undertaken with humility and wonder. This is the Sacred Scientific Method — the first and most enduring science of all.

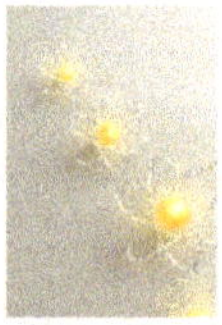

Living Science

Chapter 2

The Holy Blunder

Discovery Born From Mistake

Witness

In 1928, Alexander Fleming returned from vacation to his laboratory. On his desk sat culture plates of Staphylococcus bacteria he had forgotten to clean. By accident, a mold spore had drifted onto one plate and grown. Around the mold, the bacteria had been destroyed, leaving a clear ring.

Many scientists would have discarded the plate as ruined. Fleming paused, studied the oddity, and realized that something powerful was at work. That mold — Penicillium — would become the first widely used antibiotic, penicillin.

Reflection

The divine signature lies not in perfection, but in blunder. What most would call a mistake or contamination, became the seed of healing for millions.

The wide arc of creation bends to use even what we overlook. Fleming's forgotten plate teaches us that providence can work through neglect, accident, and even error. So it is with us: the missteps, the overlooked corners of our lives, the places we meant to discard — these may be the very spaces where grace is at work, growing rings of healing wider than we can see.

Sacred Margin

Recall a time when what you thought was failure later bore unexpected fruit. Rest with the truth that accidents may be divine appointments in disguise.

Charge

Do not despise the blunder.

Trust that even your forgotten plates may grow medicines for the world.

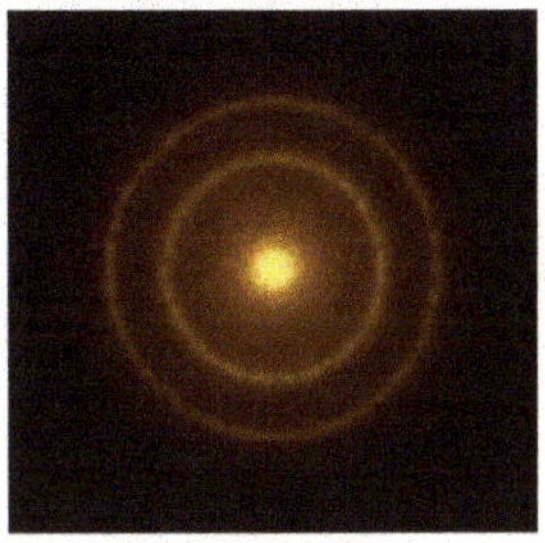

Living Science

Chapter 3

The Nested Design

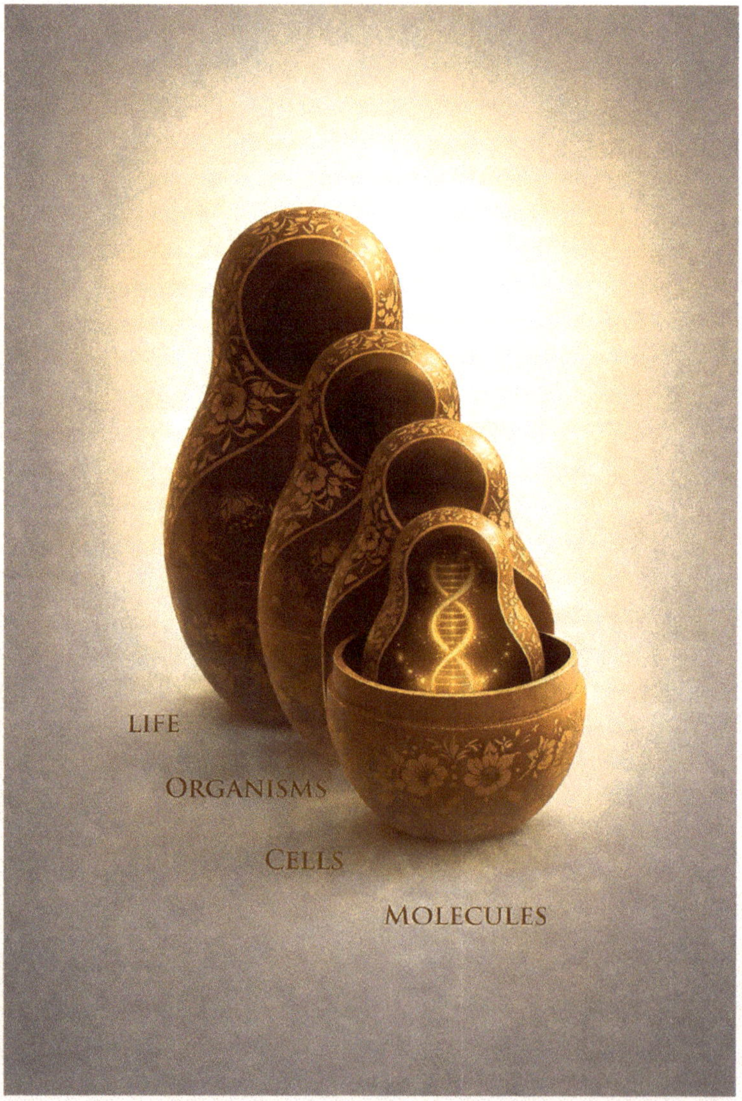

Layers of Life & DNA

Witness — The Nested Design

My father used to ask, "How do you eat an elephant?" The answer: "One bite at a time."

It was his way of saying that no matter how large the task, the only way forward is to break it down into pieces.

In the classroom, I carried that same spirit. When we want to understand something as immense as life itself, the best path is to break it down into smaller parts. Science names these the **levels of organization**: atom → molecule → cell → tissue → organ → organism → population → ecosystem → biosphere.

I imagine them like Russian dolls, each one nesting inside the other. The smallest, the atom, almost disappears in your hand. Layer by layer they grow — molecules, cells, organs, organisms — until at last the outermost doll stands tall, painted with the face of the Earth herself.

Every doll has meaning on its own, yet none exists apart from the others. Each is a vessel, a womb, a shelter for what is inside.

This is not only biology. This is creation. This is design.

Confluencism sees in this nesting an ancient truth: the Source dwells in every layer. The divine spark is not only at the grand outer shell but also hidden in the innermost seed. What looks like separation is in fact unity at work — a chorus of scales, from the very small to the very great.

And so, the elephant and the dolls carry the same message: The whole cannot be swallowed in one gulp, nor can it be understood by staring only at the outside. Wisdom comes "one bite at a time," one layer at a time, but always with the knowledge that every bite belongs to the same body, and every layer belongs to the same song of life.

Living Practice

Pause today and notice the "dolls" you belong to. See yourself as an individual, yes — but also as part of your family, your community, your people, your Earth.

Honor the smallest, honor the largest, and honor the way they fit together. In this practice, unity is no longer an idea but a rhythm you can live.

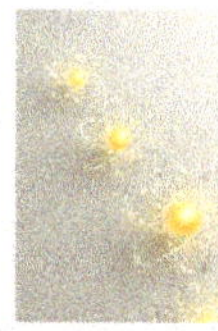

Living Science

Chapter 4

DNA: The Written Word In Flesh

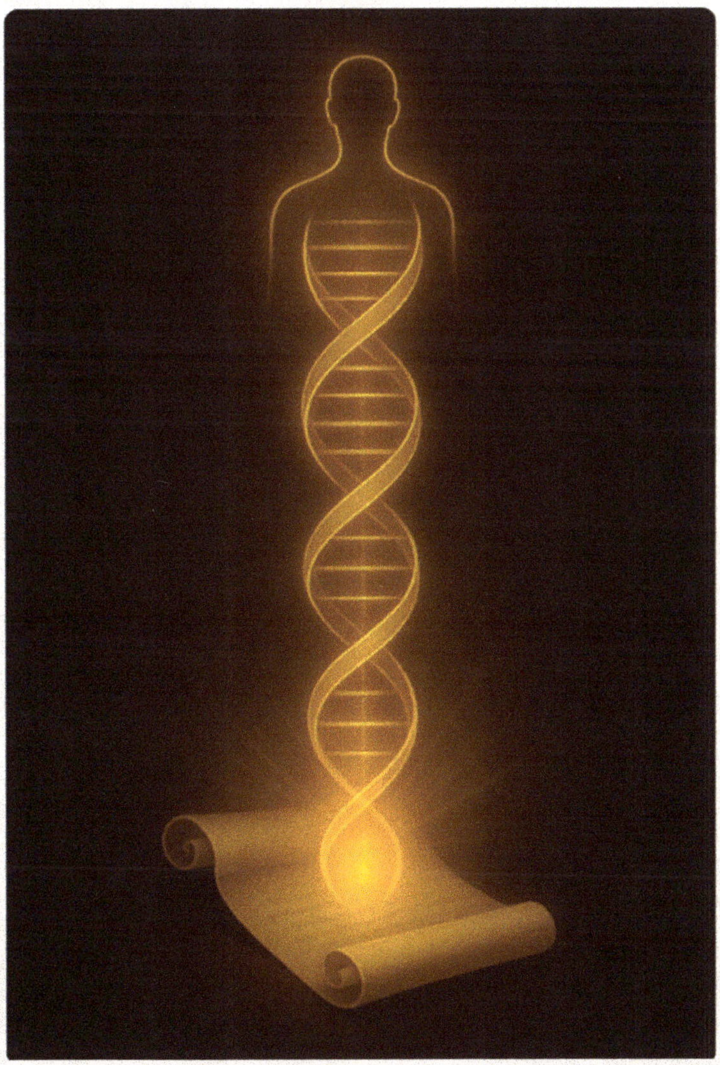

The Written Code of Life

Preface — The Written Word in Flesh

Every culture treasures sacred text.

Ink on parchment, words engraved on stone, songs passed down through memory — each carries meaning across generations.

In creation, **DNA is the sacred script written into every living being.** It is the Word in flesh, carrying the story of life itself.

Witness — The Code of Life

"The code of life is written in every cell, a living scripture passed from generation to generation."

DNA is the genetic code — four simple letters (A, T, C, G) composing the instructions for every living organism.

This code is both universal and particular:

- **Universal in structure** — all life shares the same alphabet.
- **Particular in expression** — the same letters write endless varieties of life.

Spiritually, DNA teaches:

- **Unity of design** — creation shares one language.
- **Identity and diversity** — infinite uniqueness from the same four bases.
- **Transmission** — life carried forward faithfully through generations.

Integrated Witness Panel

Canonical

- "The Word became flesh and dwelt among us." (John 1:14)
- "Your word is a lamp to my feet." (Psalm 119:105)

Cultural

- Hebrew scrolls as carriers of the covenant
- African oral tradition — stories as lifeblood of community

Scientific

- DNA as the universal genetic language
- Human genome: 99.9% identical, with differences woven into sameness

Modern

» Digital code echoing biological code
» Family ancestry tests reconnecting people to their heritage

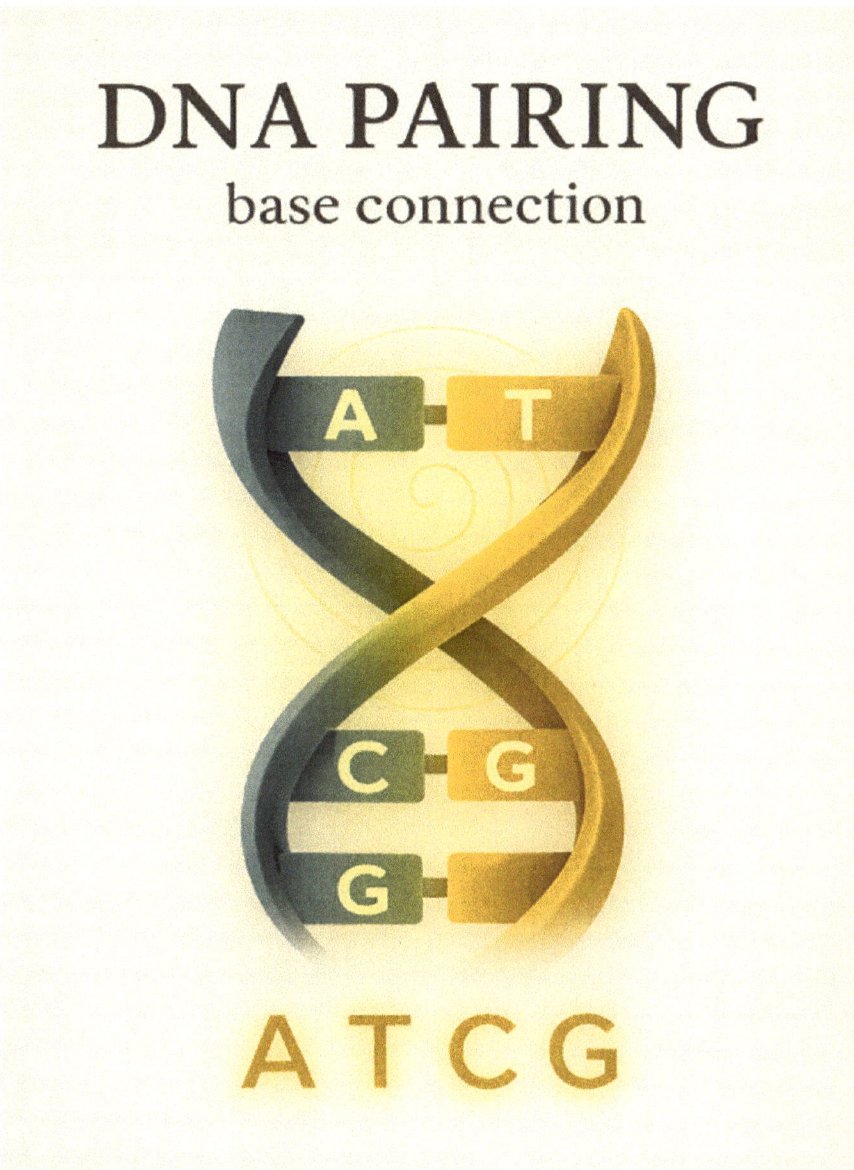

Reflection

DNA is both story and structure. It tells us that creation is inscribed with meaning, and that every person is written from the same alphabet of life.

Your DNA is ancient — a script carried through countless generations — and yet your expression of it is entirely new.

The Word is alive in all creation.

Sacred Margin

Close your eyes.

Imagine every cell in your body holding a library of sacred text.

Consider:

What story does my life write with this script?

How do I honor the Word in others?

Charge

Treat life as sacred text.

Honor the code written in all creation.

For in every being, **the Word is alive.**

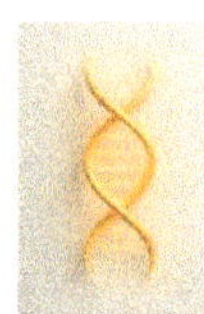

Living Science

Chapter 5

The Instigators: Enzymes And The Energy Of Change

Cradling the Spark of Energy

Witness — Enzymes and the Energy of Change

In every living cell, countless chemical reactions must occur. But most of them would take far too long on their own. Left unaided, life's chemistry would stall.

Enter **enzymes** — proteins that act as catalysts. They lower the "activation energy," the barrier that keeps molecules from reacting. Enzymes don't become part of the product; they simply make the meeting possible, speeding the process, nudging molecules together.

Without enzymes, digestion, respiration, and DNA replication would crawl at a pace incompatible with life. With enzymes, life becomes a constant flow of transformation — reactions rippling one into another.

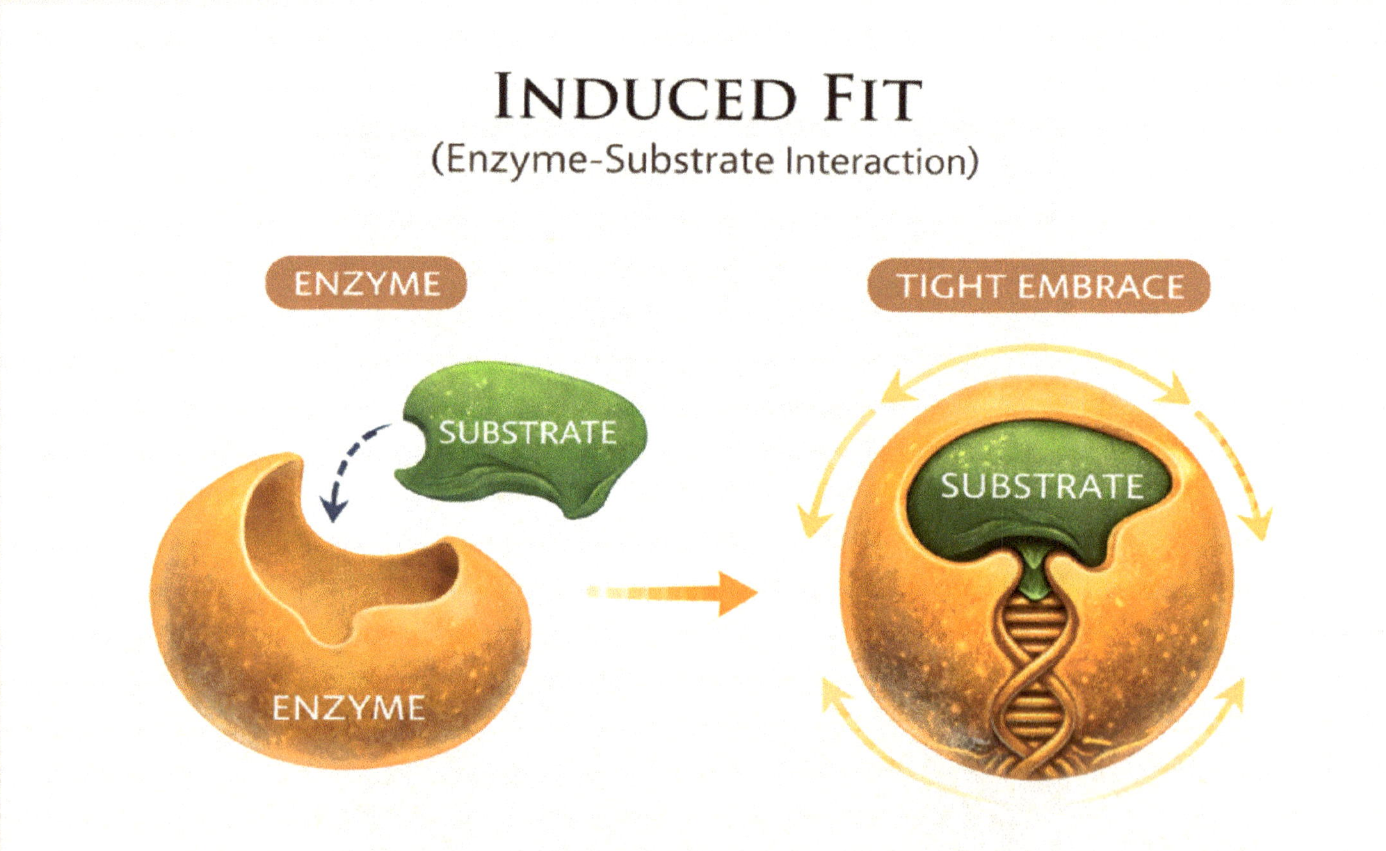

Reflection

Enzymes teach us that change often needs a mediator.

Molecules may sit side by side for ages, unable to react, until something steps in to make the connection. The enzyme does not force — it lowers the barrier. It creates the conditions for a reaction to happen.

This is a sermon about community and spirit. Sometimes we, like molecules, sit locked in a standstill — potential present but unrealized. Then a mediator enters: a word spoken, a teacher guiding, a friend reconciling, the Spirit stirring. Suddenly, the reaction begins. Transformation takes place.

Enzymes also teach us about power in humility. They do not dominate or become the product. They work quietly, unseen, then step aside, leaving life changed. True mediation is service, not self-glory.

Sacred Stream

Paul writes: "There is one mediator between God and mankind, the man Christ Jesus." (1 Timothy 2:5)

Biology writes: There is one catalyst between molecules — the enzyme that lowers the barrier to transformation.

Together they declare: Life depends on mediation. Barriers too high for us alone can be lowered by one who serves as a bridge. Whether in cell or spirit, **change requires a catalyst.**

Living Science

Chapter 6

The Smallest Shepherds — Guardians Below The Cell Level

Guardians of Balance Unseen

Sacred Witness

"For behold, I will bring health and cure, and I will cure them, and will reveal unto them the abundance of peace and truth." — Jeremiah 33:6

The ancient promise of health is not only the absence of disease, but the unveiling of peace and truth within creation itself.

Hidden in the soil, the rivers, and even within our own bodies are unseen guardians — reminders that healing is woven into the design of the world.

Cultural Stream

In African wisdom, **ubuntu** teaches: "I am because we are."

The strength of the community depends on the balance of the whole. A single sickness — in the body or in the village — calls for a careful response, not reckless destruction.

Healing means restoring the one without harming the many.

Scientific Insight

Bacteriophages — the most abundant biological entities on Earth — dwell in soil, oceans, rivers, sewage, and the human gut.

Each is a precise hunter, designed not to wipe out entire bacterial kingdoms but to attach only to a single target.

They are living parables: **invisible shepherds correcting imbalance through specificity.**

As bacteria evolve resistance, phages evolve in response. Creation itself built resilience into the smallest scale of life, offering cures from the very ground we walk on.

Living Practice

Discernment: When facing overwhelming problems, resist the urge to scorch the earth. Ask: What is the precise point of imbalance that needs my attention?

Humility: Just as phages dwell in the same soil as the bacteria they balance, expect that solutions often live near the problem. Look first to the "neighborhood" of the struggle.

Adaptation: Allow your responses to grow as challenges grow. Phages do not quit when resistance arises — they adapt, and so can we.

Confluence Reflection

Bacteriophages teach us that balance in creation is not kept by power alone, but by **relationship, precision, and patience**.

They are the smallest shepherds, reminding us that even in the invisible places, **abundance of peace and truth is already at work.**

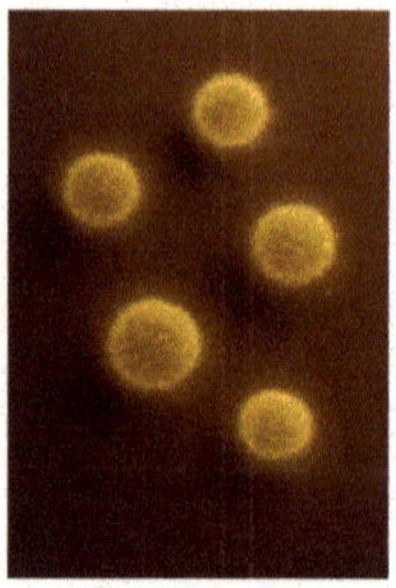

Living Science

Chapter 7

The Sacred Ratio: Why Cells Stay Small

Golden Cubes Above Blooming Lotuses

Witness — Why Cells Stay Small

Every living thing is built from cells. But cells are not giant blobs stretching endlessly. They are tiny, microscopic units. Why? Because **life depends on exchange.**

For a cell to live:

» nutrients must enter
» wastes must exit
» gases must diffuse
» signals must pass

The efficiency of this exchange is determined by the **surface area–to–volume ratio (SA:V).**

Here is the paradox: as a cell grows larger, its volume increases much faster than its surface area. Double its size, and the interior grows eightfold while the surface grows only fourfold.

Soon, the interior cannot be fed, and waste cannot escape fast enough.

This is why cells remain small. Small cells possess a greater surface area relative to their volume, giving them efficiency. And when life needs something larger, it does **not** make one giant cell — it makes **many small ones working together in community.**

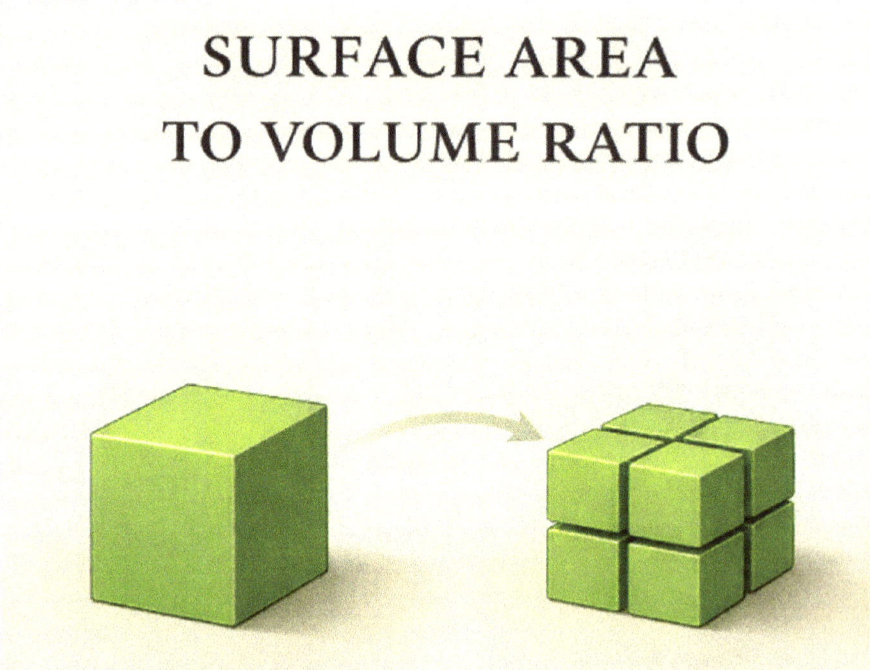

Reflection

The sacred ratio teaches that greatness is sustained by smallness.

The vastness of an oak tree, a whale, or a human being depends on the faithfulness of trillions of tiny cells.

It also teaches that **openness is life.** A cell must remain porous, exchanging with its surroundings. If it swells too large, too self-contained, it starves. So too with us: when we become closed off, wrapped in ourselves, or isolated, the flow of life is cut off.

And it teaches the wisdom of multiplicity. Life thrives not in one grand structure, but in countless small units, each open to exchange. Community, diversity, cooperation — these are not human inventions but the very architecture of biology.

Sacred Stream

Jesus said, "The kingdom of heaven is like a mustard seed… though it is the smallest of all seeds, yet when it grows, it is the largest of garden plants." (Matthew 13:31–32)

Biology says: The cell is small, yet from it arise forests, oceans, and civilizations.

Together they declare: **Smallness is not weakness but wisdom.** The sacred ratio whispers that life flourishes not by swelling in isolation, but by staying small enough to remain open, and connected enough to belong.

Living Science

Chapter 8

Osmosis: Balance & Spirit Flow

A Tree Within a Glowing Droplet

Preface — Balance & Spirit Flow

Water moves toward balance. Through **osmosis**, it flows from where it is abundant to where it is scarce, bringing equilibrium across membranes.

This hidden current in every cell is a sacred parable: **the Spirit, too, moves wherever imbalance exists — bringing life and harmony.**

Witness — The Flow Toward Need

"The Spirit flows toward need, filling what is empty, balancing what is out of order."

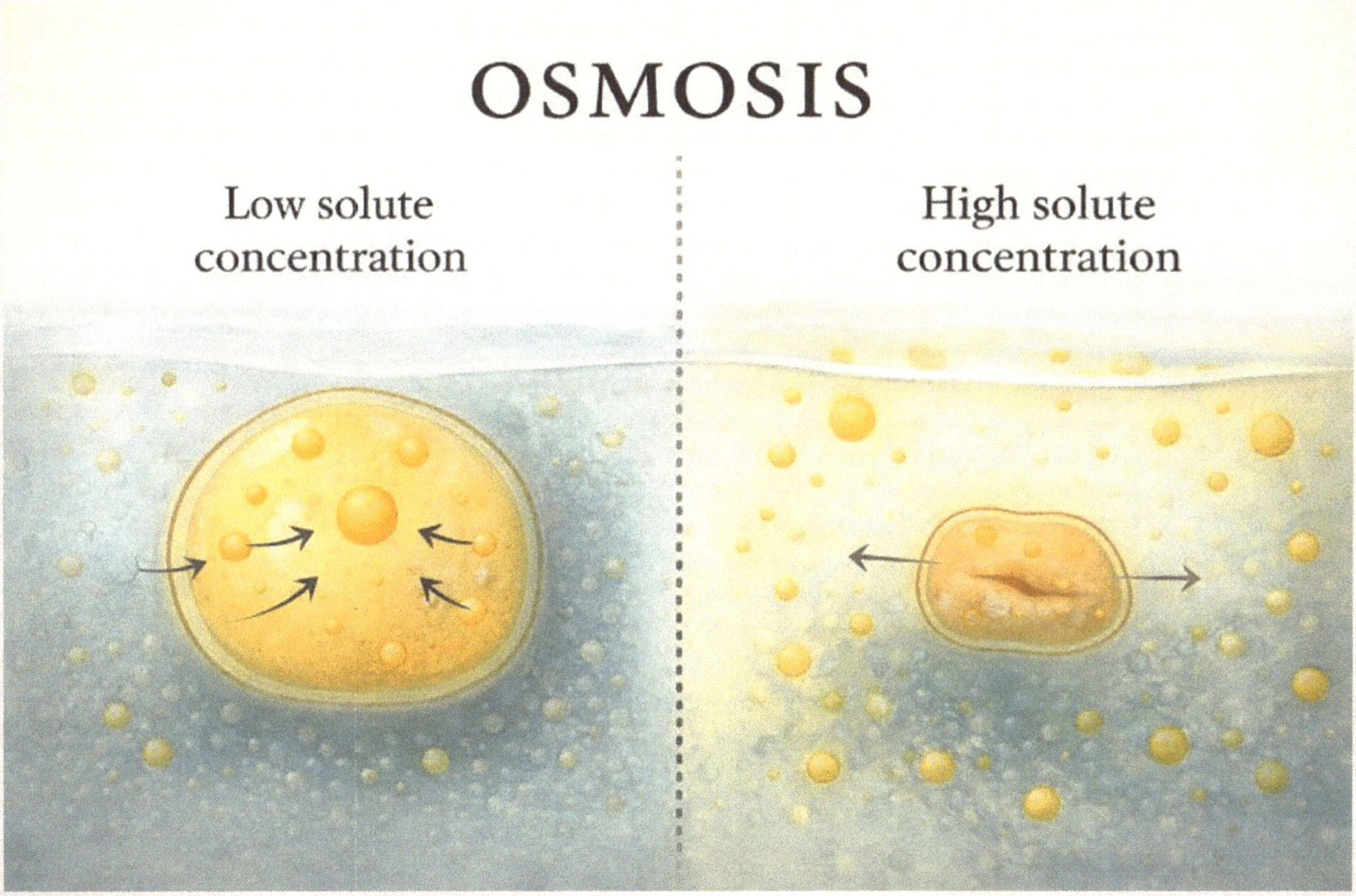

Reflection

Osmosis is the movement of water across membranes.

- Too much water inside a cell → it bursts.
- Too little → it shrivels.
- **Life is sustained when flow restores balance.**

Spiritually, osmosis teaches:

- **The Spirit moves toward need** — healing imbalance.
- **Discipline is balance** — too much or too little destroys.
- **God's flow sustains life** in the right measure.

Osmosis becomes scripture written in water — a sermon about equilibrium within and without.

Integrated Witness Panel

Canonical

- "Whoever believes in me, rivers of living water will flow from within them." (John 7:38)
- "He leads me beside still waters." (Psalm 23)

Cultural

- African proverb: "When the rhythm of the drum changes, the dance also changes."
- Taoist teaching: water adapts to every vessel, yet wears down stone.

Scientific

- Osmosis maintains cellular stability.
- Water flow across membranes is essential to life.

Modern

- Emotional balance — too much or too little consumes.
- Social balance — justice flows toward what is lacking.

Sacred Margin

Imagine your soul as a vessel.

Where are you swollen with excess?

Where are you shriveled with lack?

Ask:

Where do I need Spirit's balancing flow?

How can I let life's currents move freely through me?

Charge

Seek balance. Let Spirit flow where you are empty, and restrain where you overflow.

For life is sustained by **balance**.

Living Science

Chapter 9

Stem Cells: Seeds Of Potential

Seeds of Cellular Potential

Witness — The Stem Cell and the Power of Potential

In the intricate design of living biology, the stem cell stands as a symbol of pure potential. A stem cell is a blank slate, capable of becoming any number of specialized cells depending on the guidance it receives.

It is the ultimate biological reminder that life is filled with transformative possibility.

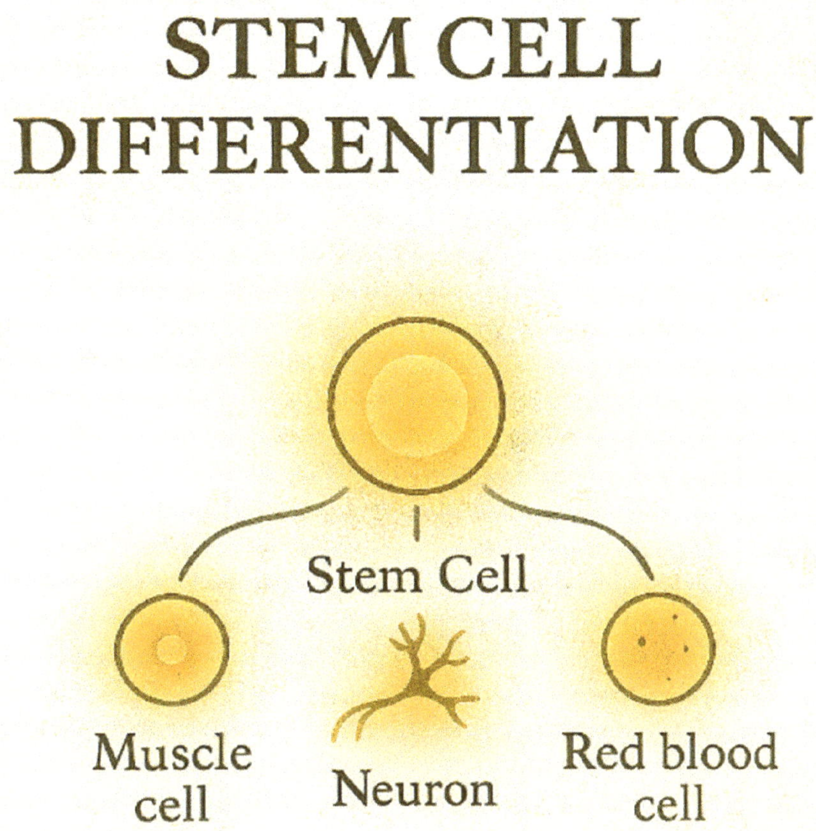

Reflection

Just as a stem cell can become a muscle cell, a nerve cell, or a blood cell, we too have the capacity to transform and adapt on our journey.

Our environment, our influences, and the guidance we embrace shape what we become.

The stem cell reminds us that we are never locked into one form; we are always capable of growth and new possibilities.

Sacred Margin

Take a moment to reflect on an area of your life where you feel stuck or limited. Imagine yourself as a stem cell, holding the potential to grow into something new with the right nurturing.

Charge

Today, embrace **one small step** of transformation. Like a stem cell responding to its environment, allow yourself to adapt and grow into the person you are becoming. Each step is a testament to the sacred potential within you.

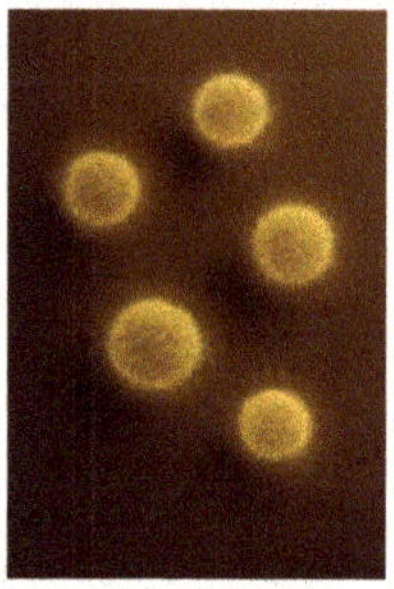

Living Science

CHAPTER 10

Crossing Over: The Sacred Exchange

Intertwining Light Between Hands

Witness — The Sacred Exchange

In the making of human life, ordinary cell division will not do. To form eggs and sperm, the body uses a special process called **meiosis** — a "reduction division" that halves the chromosome number so that two cells can join in fertilization.

But in the middle of meiosis I, something remarkable takes place. Chromosomes from the mother's line and chromosomes from the father's line pair up. Instead of staying separate, they trade pieces of themselves. This process is called **crossing over**.

Segments of DNA are exchanged, blending traits from both lines into unique combinations.

The outcome is astonishing: no two sperm, no two eggs, no two children are the same. Each carries both continuity with the past and a new arrangement for the future.

Reflection

Crossing over is creation's way of preserving both **memory and possibility**. It ensures that generations are not broken but carried forward, strand by strand. Yet it also ensures that life never repeats itself exactly.

Each person is kin, but each person is also unique.

Spiritually, this moment whispers of the sacred pattern of exchange: the mingling of differences that brings forth life. It is not sameness that keeps creation alive but **variety**.

Each crossing preserves lineage while guaranteeing novelty. Each crossing says: you are part of the story, but you are also something never seen before.

This is also a lesson in humility. None of us is self-made. We are made by the crossing of lines, the weaving of stories, and the exchange of gifts passed down from ancestors.

We carry in our bodies a sermon that no generation stands alone.

Sacred Stream

Genesis affirms: "God created mankind in his own image… male and female he created them." (Genesis 1:27)

Biology affirms: In meiosis, images are crossed over and renewed, so that no single image repeats.

Together they proclaim: Creation thrives by exchange. Diversity is not an accident but a design. Life is strongest when lines meet, when differences cross, when inheritance and possibility join hands.

Crossing over preaches this truth: We are threads of continuity and sparks of uniqueness. The sacred exchange in our cells is also the sacred exchange of community — each of us carrying something old, each of us bringing something new.

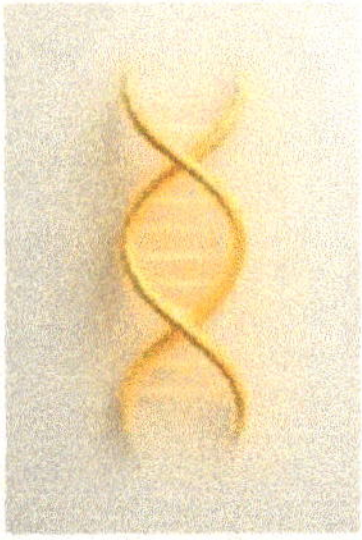

Living Science

Chapter 11

Genetic Paradoxes: The Hidden Strength Of Weakness

Butterfly & Thorns

Witness — The Hidden Strength of Weakness

Across the world, there are traits in human DNA that appear at first to be curses.

» **Sickle cell disease** twists red blood cells into sharp crescents.

» **Thalassemia** leaves blood struggling to carry oxygen.

» **Tay-Sachs** and **cystic fibrosis** shorten lives.

But look closer, and creation tells another story.

In Africa, India, and the Mediterranean, carriers of the **sickle cell trait** are resistant to malaria. What looks like a flaw in one place is a shield in another.

In southern Europe and the Middle East, **thalassemia traits** also guard against malaria, keeping populations alive where the disease once raged.

Among Ashkenazi Jewish communities, **Tay-Sachs carriers** may have had protection against tuberculosis in crowded ghettos.

In Europe, **cystic fibrosis carriers** may have been more resistant to cholera and typhoid, their altered salt balance sparing them from severe dehydration.

These "disorders" were not random cruelties. They were part of the dance between human life and its environment — **survival etched into bloodlines by struggle.**

Reflection

This is a paradox: what harms in one form helps in another. What seems like brokenness may hold hidden wisdom. Creation writes resilience into weakness, strength into suffering.

This truth humbles us. We cannot label life too quickly as a curse or a blessing. A trait that devastates one generation may have saved countless others.

What we call disorder may be creation's strategy for survival.

And in this, biology mirrors the story of faith:

» Weakness becomes the vessel of strength.

» Suffering becomes the seed of endurance.

» That which we would discard, God redeems.

Sacred Stream

Hebrews affirms: "Out of weakness were made strong." (Hebrews 11:34)

Paul testifies: "My grace is sufficient for you, for my power is made perfect in weakness." (2 Corinthians 12:9)

Biology testifies the same:

» What looks like a defect may be a defence.

» What looks like a burden may be a blessing.

Together they declare: **Creation's paradox is also God's truth — that life often survives not in spite of weakness, but through it.**

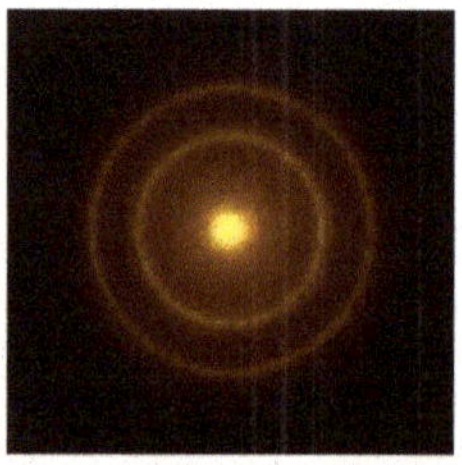

Living Science

CHAPTER 12

Strategies Of Survival: The Few And The Many

Survival Through Care & Memory

Witness — The Few and the Many

Species across creation do not all survive in the same way. Biologists describe two broad strategies:

R-strategists (from "rate" of reproduction):

Organisms like mice, insects, or weeds. They produce **many** offspring, investing little in each. Most will not survive, but sheer numbers ensure that some will.

K-strategists (from "carrying capacity"):

Organisms like elephants, whales, or humans. They produce **few** offspring, each one nurtured, protected, and guided. Their survival depends not on numbers but on care and longevity.

Neither strategy is "wrong." Each fits the environment and the call of the species.

But humans clearly belong among the K-strategists:

» We raise children slowly.

» We pass down wisdom.

» We depend on groups, families, and communities for survival.

Reflection

This difference preaches.

For some, survival is in **scattering seed far and wide**, hoping some will endure.

For others, survival is in **depth** — pouring life into the few, guiding them, teaching them, surrounding them with community.

Humanity thrives not by sheer numbers but by **wisdom**.

Our young cannot walk or run within minutes of birth like a foal or a calf. They rely on parents, on elders, on villages.

We survive through:

» Storytelling

» Guidance

» Memory

» Community

This is a sermon about the sacredness of care. We are not designed to stand alone. We are designed to be raised, nurtured, guided, and loved.

Parenting — biological or spiritual — is not optional; it is essential to our survival strategy.

Sacred Stream

Proverbs teaches: "Train up a child in the way he should go; even when he is old he will not depart from it." (Proverbs 22:6)

Biology teaches: Humans are K-strategists, relying on care and guidance rather than numbers alone.

Together they declare: **Our survival is not secured by scattering many, but by nurturing the few.**

Creation itself preaches that wisdom, guidance, and community are our way of enduring.

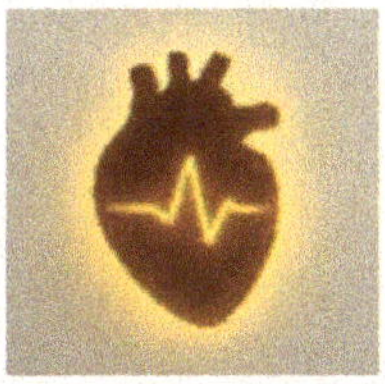

Living Science

CHAPTER 13

The Ecology Of Relationships

Ecology of Interdependence

Witness — The Ecology of Relationships

"Look at the relationships written into creation: some give and take, some give and give, and some take without giving back. These too are parables of life."

In ecology, symbiosis describes how different lives intertwine.

» **Parasitic** — one thrives at the expense of another.

» **Mutualistic** — both partners flourish.

» **Commensal** — one benefits while the other is unharmed.

Creation is teaching us something here.

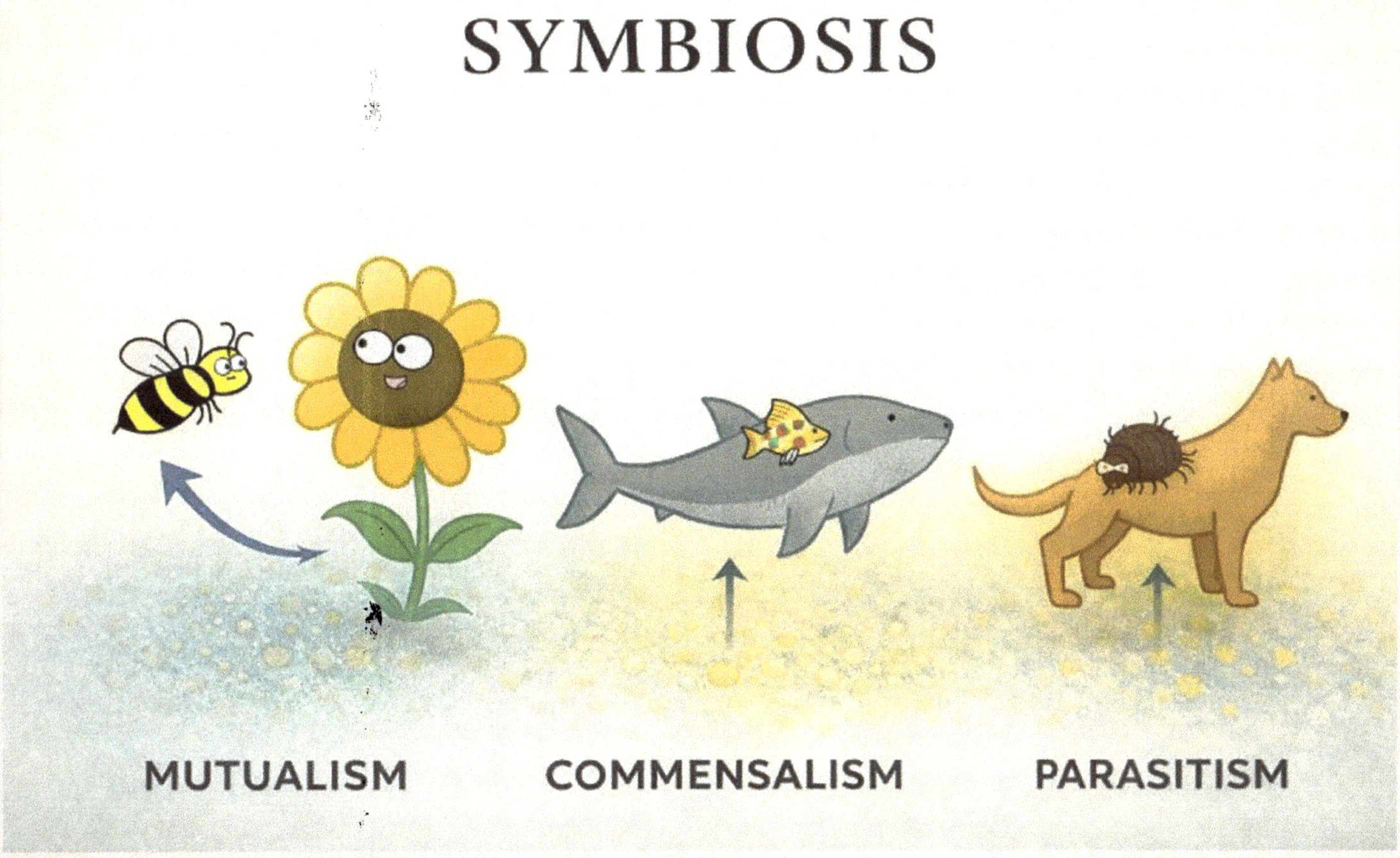

Reflection

Parasitism warns us against becoming **drainers of life** — people or patterns that sap strength while feeding themselves. Spiritually, this mirrors sin that clings, idols that promise life but slowly consume it.

Mutualism points to **covenant, community, and divine love** — friendships, families, mentorships, and even our relationship with God, where giving and receiving build life on both sides.

Commensalism reminds us of **grace**: the tree loses nothing when it shades the traveler. God is never diminished by blessing us. Sometimes we are called to receive gifts without guilt, knowing the Giver is not lessened.

The ecology of creation whispers: **examine the kind of relationship you are in — and the kind you are becoming.**

Sacred Margin

Sit quietly and recall three relationships:

1. One that feels draining.

2. One that feels mutual.

3. One where you receive freely without cost to the other.

Ask yourself:

» Where might I need to release what drains?

» Where can I deepen mutual flourishing?

» Where can I receive blessing with gratitude instead of guilt?

Charge

Step into this day with eyes open to **the ecology of your relationships**. Let creation remind you: **flourishing is found where love is rightly ordered.**

Living Science

CHAPTER 14

Seeds: Promise Encased

Promise Waiting In Silence

Preface — Promise Encased

Every seed is a prophecy of tomorrow wrapped in humility. It hides power inside silence, life inside stillness.

To the untrained eye, a seed looks like nothing more than dust— but within that husk wait forests, harvests, and generations yet to come.

Witness — The Trinity Within

In every seed, three parts conspire for survival:

» **Embryo** — dreams of form

» **Endosperm** — stores food for the journey

» **Seed Coat** — guards the treasure

Faith, provision, and protection— **biology's own trinity of hope.**

Buried beneath the soil, pressed by darkness and time, the seed obeys the invisible call to rise.

Reflection

Creation preaches resurrection through every seed.

Dormancy is not death; it is waiting with purpose.

Inside the soil's womb, enzymes awaken, coats soften, and potential stirs.

So it is with us: periods of silence, obscurity, or burial may only be seasons of gestation. The ground that covers you may be the very ground that prepares you.

The seed also teaches restraint. It does not sprout on command or in comfort— it waits for the right measure of warmth and moisture.

Faith, too, has timing. **Patience is part of the miracle.**

Integrated Witness Panel

Canonical

"Unless a grain of wheat falls into the earth and dies, it remains alone; but if it dies, it bears much fruit." (John 12:24)

Cultural

- African proverb: "He who plants a seed plants hope."
- Indigenous teaching: "The seed remembers the tree."

Scientific

- Embryo → future form
- Endosperm → stored energy
- Seed coat → defense
- Dormancy preserves life until conditions align

Modern

- Seed banks guarding biodiversity
- Communities preserving wisdom for a coming age

Sacred Margin

Hold one small seed in your hand—or picture one. Imagine the power asleep within.

Ask yourself: What vision in me lies dormant, waiting only for season and soil? What must fall, what must rest, before it can rise?

Charge

Plant something—literally or figuratively. Then guard it, feed it, wait for it.

Let patience be your act of faith. For the kingdom of God and the rhythm of life whisper the same truth: **the smallest beginnings hold the largest futures.**

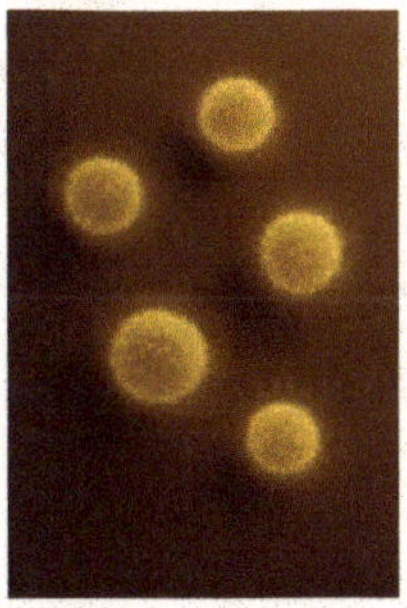

Living Science

Chapter 15

Succession: The Return Of Life

Life Returns in Stages

Witness — The Return of Life

When a forest burns, a glacier melts, or a city is abandoned, it does not remain barren forever. Creation has a way of taking ground back. This process is called **ecological succession**.

Primary succession

Begins on bare rock. Lichen and moss arrive first, breaking stone into soil. Soon grasses take hold, then shrubs, then trees— layer by layer until a whole ecosystem thrives.

Secondary succession

Occurs where soil already remains. After a fire, flood, or human departure, pioneer species move in quickly— dandelions, weeds, fast-growing plants— followed by deeper-rooted vegetation and rising complexity.

Given time, abandoned streets crack, weeds sprout, roots push through concrete, and forests reclaim the ground. What looked like ruin becomes a new sanctuary.

Reflection

Succession preaches that **life returns**. Desolation is not final. Nature is patient, resilient, and relentless. Even where humans see wasteland, creation sees **a canvas**.

It also preaches that **restoration is layered**. Renewal does not arrive all at once. It begins small— lichen on stone, weeds through cracks— then grows step by step. Humble beginnings prepare the way for greater flourishing.

And succession preaches **humility**. Humanity imagines it rules the Earth, yet if we step aside, nature moves in.

Forests will rise where towers once stood. Rivers will reclaim their banks. Birds will nest in rafters. Trees will root in sidewalks. Vines will climb steel.

We are not masters— only **tenants**.

Sacred Stream

Isaiah declares: "The wilderness and the dry land shall be glad; the desert shall rejoice and blossom like the crocus." (Isaiah 35:1)

Biology declares: Barren places will be clothed again— soil formed, plants rooted, ecosystems reborn.

Together they proclaim: **Renewal is woven into creation.** What burns will bloom again. What lies silent will sing again. Death is never the final word, for God has written resilience into the very soil.

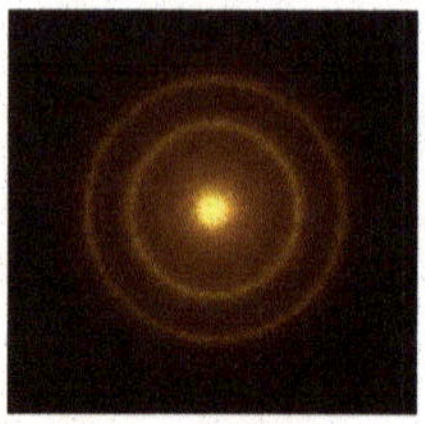

Living Science

CHAPTER 16

Milk & Mercy

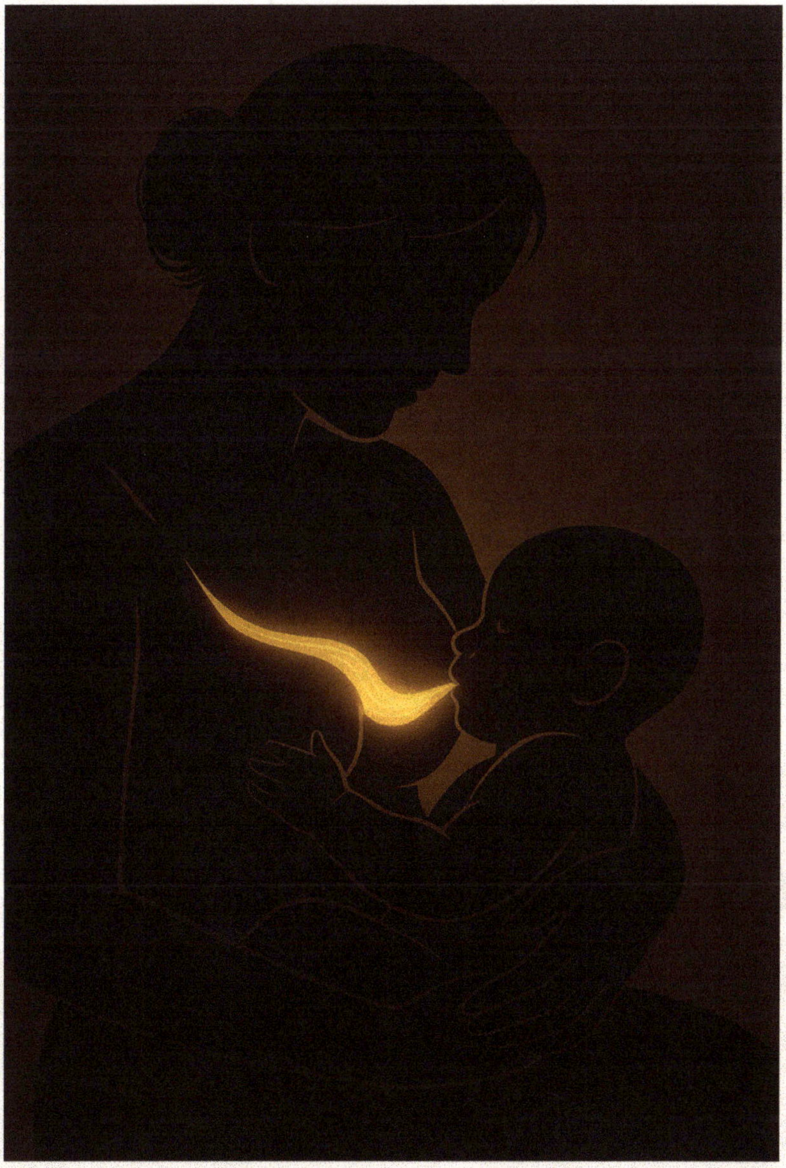

Nourishment as Care

Witness — Milk & Mercy

From the first cry of life, nourishment flows.

Breast milk — a living substance, not mere food — arrives as nature's first sacrament.

It carries:

» antibodies to guard,

» fats to warm,

» sugars to sustain,

» and proteins that teach the infant's cells what safety feels like.

This milk remembers the mother's encounters with the world; **her immune system becomes the child's inheritance.**

Each drop whispers: "You are not alone; your needs are known before you can name them."

Reflection

Here, **mercy becomes molecular**.

The mother's body translates compassion into chemistry — **mercy made measurable.**

Before words or prayer, before doctrine or discipline, **love manifests as flow.**

In Confluencism, this is the first communion of creation: life feeding life — not for profit or merit, but because **nurture is divine law**.

Just as milk adapts to a child's need — richer after illness, lighter in heat — so does mercy meet us where we are.

We do not earn it; we are nourished by it.

To be human is to both receive and extend this milk of mercy — to let compassion move through us as instinct, not an obligation.

Integrated Witness Panel

Canonical

"As a mother comforts her child, so I will comfort you." — Isaiah 66:13

Cultural

African proverb: "The breast does not forget the mouth that first drew from it."

Scientific

Breast milk contains adaptive antibodies, hormones, and growth factors that **change with each feeding**, responding to the infant's need.

Modern

Donor milk banks, neonatal care, and community feeding programs echo the same principle — **shared nurture sustains life.**

Sacred Margin

When mercy stops flowing, the spirit of creation begins to hunger.

When it flows freely, the world heals.

As milk carries life from one body to another, forgiveness carries peace from one soul to another.

Let compassion move through you as the milk of mercy moves through creation — **sustaining, adapting, never running dry.**

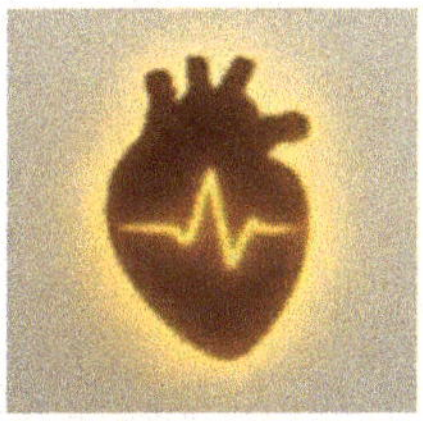

Living Science

Chapter 17

Surface Treasures Of The Earth

Gifts Found Without Digging

Witness — Surface Treasures

"The earth is full of the goodness of the Lord." (Psalm 33:5)

The earth hides most of its wealth in deep places, but sometimes it lays treasures right at the surface.

» Hot springs bubble up carrying minerals, leaving terraces of color and stone.

» Salt domes rise until they break open — a spring of salt for anyone to gather.

» Oil and tar ooze from the ground as if the earth itself were bleeding richness.

» Rivers wash loose gold into streambeds, where prospectors can find nuggets without digging.

» Native silver gleams as streaks in exposed rock.

These are **surface treasures** — gifts creation places at our feet.

Reflection

Surface treasures remind us that not every gift must be earned through struggle.

Creation offers both:

» the labor of mining, and

» the grace of discovery.

Where deep treasures require excavation, surface treasures teach **receptivity** — the art of noticing what is already offered.

In the economy of grace, both patience and awareness are sacred virtues:

» The deep teaches endurance.

» The surface teaches gratitude.

Not every blessing comes through digging. Some blessings simply appear. The spiritual work is to recognize them.

Integrated Witness Panel

Canonical

"The earth is full of the goodness of the Lord." (Psalm 33:5)

Cultural

Indigenous wisdom: "The earth gives what is needed to those who walk gently."

Scientific

Surface mineral deposits form through:

» hydrothermal venting,

» erosion,

» plate movement — creation's way of revealing wealth **without destruction**.

Modern

Renewable energy — sunlight and wind — are today's "surface treasures," resources drawn from the **top** of the world rather than its wounds.

Sacred Margin

As you walk today, notice the surface treasures that appear without your digging —

» an unexpected kindness,

» a word of encouragement,

» a door opened.

Gather them with gratitude. They are the Creator's way of saying: **"See, I have made this easy for you."**

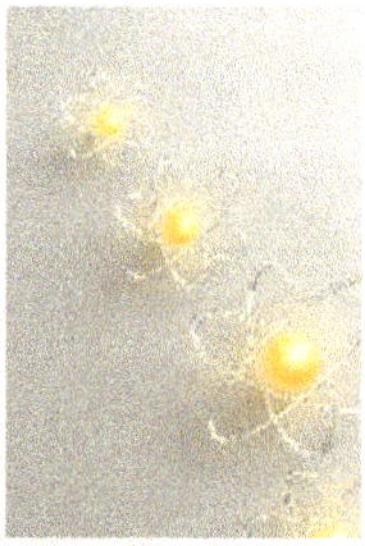

Living Science

Chapter 18

Buried Treasures Of The Earth

Hidden Treasures Beneath the Surface

Witness — Buried Treasures

"For everything there is a season, and a time for every matter under heaven." (Ecclesiastes 3:1)

Not all treasures wait at the surface. Many are born in silence and pressure, hidden for ages until the right time.

» **Diamonds** are carbon transformed under crushing weight, carried upward only by ancient volcanic fire.

» **Oil and coal** are sunlight stored in forests and swamps, compressed over millions of years until they became fuel.

» **Veins of gold and silver** formed when molten rock cooled, locked deep in the mountain's bones until uncovered by erosion.

» Even **soil** is a slow gift — centuries of rock weathering into the ground that grows our food.

These treasures cannot be hurried. They testify that creation's richness is not only sudden but also slow. What feels like scarcity now may be abundance ripening in secret.

Reflection

Buried treasures preach that **formation takes time**.

They remind us that:

» Pressure can transform,

» Darkness can protect,

» Waiting can be generative.

A diamond's brilliance begins in crushing silence. Gold's beauty begins deep in the mountain's spine. Soil's fertility begins as stone surrendering to time.

What looks hidden, delayed, or denied may simply be **forming.**

The earth teaches: **Some blessings arrive slowly because they are being shaped slowly.**

What is buried is not forgotten. It is becoming.

Sacred Margin

Imagine holding a diamond in your hand. It is small, but its story is vast — carbon breathing in the dark for eons.

Sit in silence and let patience become holy within you.

Ask yourself:

- What in my life is forming unseen?
- What treasure might be ripening in secret?
- What needs time rather than force?

Charge

Practice endurance. When the blessing you seek does not appear quickly, trust that some treasures require time.

Just as the earth hides diamonds in its depths, the Creator hides strength and wisdom in your waiting.

Be patient — **what is buried will rise in its time.**

Living Science

Chapter 19

The Energy That Is Never Wasted

Golden Light of the Forest Floor

Witness — The Energy That Is Never Wasted

Life builds itself in tiers — a **pyramid of energy**.

At the base, green plants stretch toward the sun, capturing light and weaving it into sugars.

Herbivores graze, then carnivores hunt, each step **higher yet narrower**.

Only **a tenth** of the energy rises to the next level; the rest radiates away as heat.

To the eye of efficiency, it looks like a loss. But in truth, that heat stirs winds, shapes climates, and sustains the very conditions that allow the pyramid to stand.

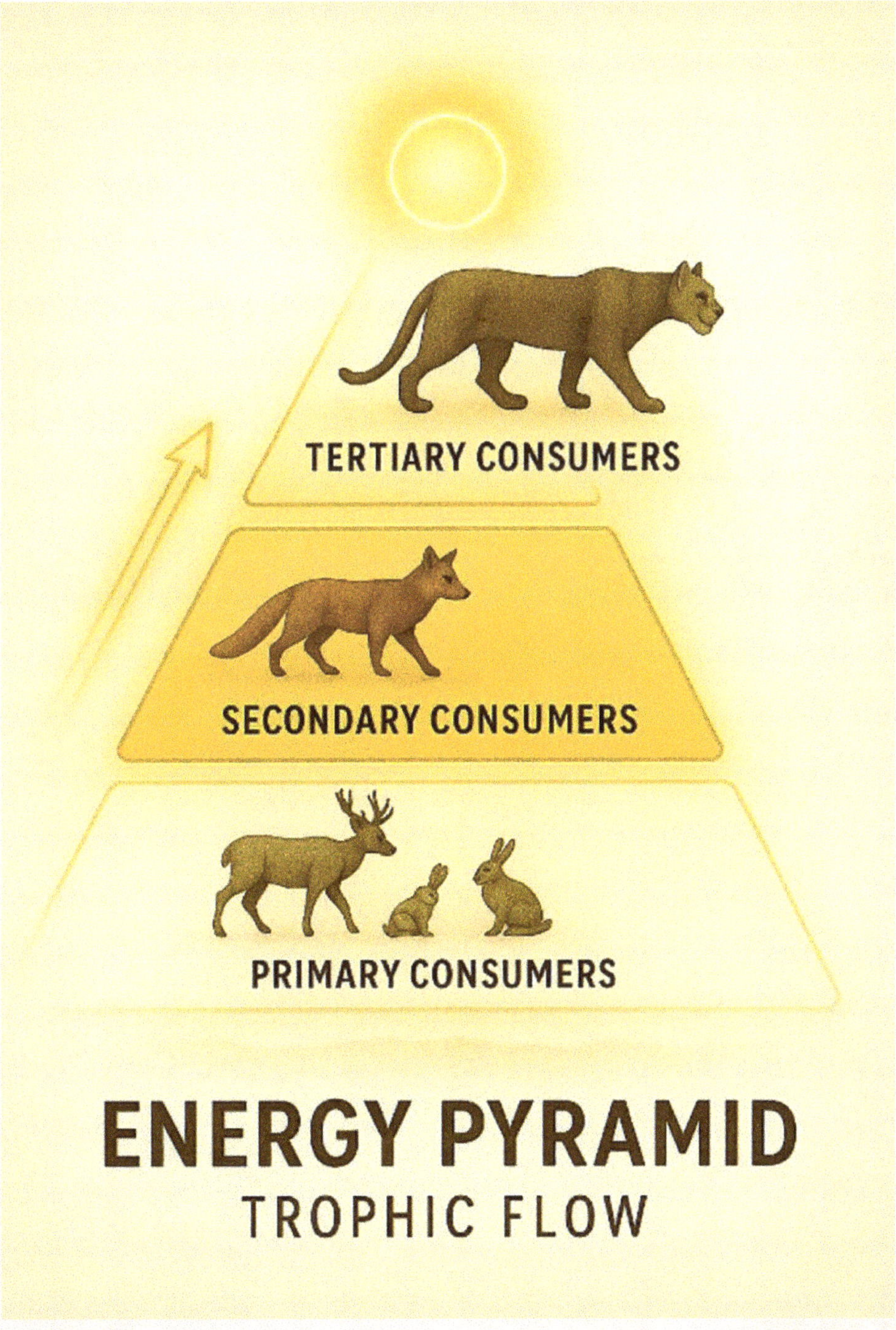

Reflection

The divine signature lies in this: **nothing is wasted.**

Energy changes form, flowing into wider blessings.

The pyramid is not stingy — it is wise.

» Its **broad base** teaches humility: the many lifting the few.

» Its **narrowing top** teaches restraint: a limit against endless consumption.

» Its **constant heat** teaches grace: what escapes our grasp still serves the whole.

So it is with us.

Our labors, our loves, even our losses — they do not vanish.

They transfigure into the hidden fabric of life, keeping balance, warming hearts unseen, holding creation in harmony.

Sacred Margin

Recall something you gave that seemed to dissolve into nothing. Rest in the truth that it still circulates.

In the divine economy, **no offering of life, love, or labor is ever wasted.**

Charge

Give freely. Accept limits. Trust the flow.

What passes beyond you is not loss, but **gift to the greater balance**.

Living Science

Chapter 20

Bulbs: The Hidden Life Beneath

Life Waiting Beneath Winter

Witness — The Hidden Life Beneath

Winter buries what spring will need.

Tulip, daffodil, lily — each hides its promise under the frost, storing life in silence.

A bulb is not dead; **it waits.**

Within its layered skin are roots and bloom, leaf and color, all folded tight until warmth returns.

It holds everything it needs to rise again.

Reflection

Not every miracle shouts.

Some are buried by design — hidden not to be forgotten, but to be protected.

The bulb teaches patience in the dark: that retreat is not defeat, and rest is not ruin.

In the soil's cool hush, it gathers sugars and strength, guided by a rhythm older than memory.

Even when no light touches it, the blueprint of beauty remains intact.

So too in us — seasons come when the surface seems barren, yet below, new faith forms.

The Creator tends us unseen, rooting renewal in quiet places.

Sacred Margin

Before you curse your winter, remember the bulb.

What is buried is not gone; **it is becoming.**

What feels hidden may be healing.

Let the frost have its hour — for mercy, like spring, will find its way through frozen ground.

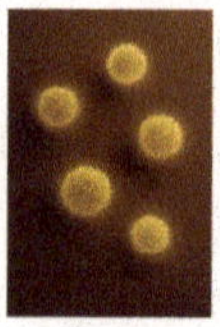

Living Science

Chapter 21

Decomposers: The Sacred Cleaners Of Creation

The Work of the Decomposers

Witness — The Sacred Cleaners

When life ends, the cleaners come.

Fungi weave through fallen wood, releasing hidden nutrients.

Bacteria bloom where decay begins, breaking down what was once proud and living.

Vultures circle; worms tunnel; the invisible and the humble return matter to meaning.

Nothing is wasted. What we call rot, **the earth calls rebirth.**

Reflection

In the great economy of creation, even **death has a vocation**.

Decomposers serve as **priests of the soil** — dismantling the old so that the new may grow.

They are mercy embodied in biology, ensuring that failure and falling apart become fertile again.

We recoil from decay, yet every forest depends on it. No blossom would rise without what the fungus frees.

In Confluencism, this is the gospel according to the ground: **Grace is the cycle that refuses to end, even at death.**

To decompose is to offer oneself back to the chorus of life.

Sacred Margin

Look closely at the forest floor.

Each leaf returning to dust is a hymn. Each fungus a quiet choir.

We, too, must learn this mercy — to let what no longer serves be broken down, so that goodness can grow again.

Bless the decomposers — **the sacred cleaners of creation.**

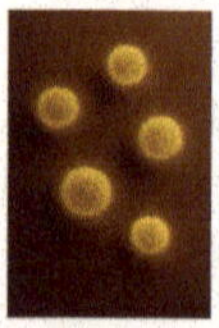

Living Science

Chapter 22

Death & Renewal: Holy Recycling

Life Emerging From Decay

Witness — Death & Renewal

Every forest must fall to rise again.

Old trees collapse, feeding the soil that will cradle their children.

Cells die by design, clearing space for new tissue to grow.

Stars burn out, scattering the elements that form new worlds.

Nothing in creation truly disappears — it only changes form.

Death is not the opposite of life; it is its sacred partner.

Together, they compose the rhythm that keeps the universe breathing.

Reflection

We fear endings because we see only **the closing of form**, not the continuity of substance.

Yet every ending is an act of generosity:

» atoms given back,

» energy released,

» spirit returned.

In the body, **programmed cell death — apoptosis —** prevents chaos; without it, growth becomes cancer.

In the ecosystem, **decay feeds birth**; without it, fertility fades.

Even in the heart, **letting go is the precondition for renewal**.

Confluencism calls this **the holy exchange — the covenant of return**.

Everything borrowed must one day be given back, not as punishment, but as participation in the eternal cycle of mercy.

Sacred Margin

Grieve what passes, but do not curse it.

The Creator wastes nothing.

The seed must crack, the leaf must fall, the tide must ebb —so that the song of life may begin again.

Where death has walked, **renewal is already sowing.**

Closing Note — Toward the Next Volume

Life does not end with understanding. It expands with attention.

Let this volume open your eyes to the divine rhythms written into cells, forests, oceans, and the quiet workings of your own body.

The next volume of Living Biology will continue the journey — exploring ecosystems, evolution, behavior, and the deeper confluence of science and spirit in the unfolding story of creation.

May your learning be worship, and your wonder be a doorway to life.

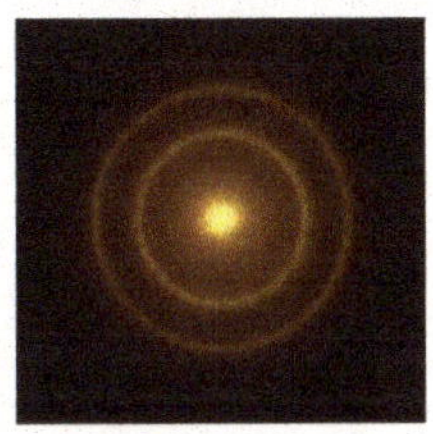

Living Science

GLOSSARY

A Confluencist Reference of Scientific & Spiritual Terms

A

Activation Energy

The energy barrier that must be overcome for a chemical reaction to begin. Enzymes lower this barrier, enabling change.

Adaptation

A biological adjustment to environmental pressure; spiritually, the ability to grow with life's challenges.

Apoptosis

Programmed cell death; the body's orderly clearing of old cells to protect life and make room for growth.

Atom

The smallest unit of matter, the foundational "seed" of physical creation.

B

Bacteriophage (Phage)

A virus that infects bacteria with precision. Confluencist metaphor: "the smallest shepherds" maintaining balance at microscopic levels.

Biosphere

All ecosystems on Earth together — the full household of life.

Biological Community

A group of interacting species living in the same environment.

Biosignature

Any chemical or physical indicator of life's presence or activity.

Bulb

An underground storage organ of plants that holds potential through winter, symbolizing hidden life.

Buried Treasures

Deep-earth resources (diamonds, coal, oil, metals) formed slowly under pressure; spiritual illustration of slow-forming blessings.

C

Carbon Cycle

The movement of carbon through organisms, soil, oceans, and air — creation's recycling system.

Cell

The basic unit of life; the smallest structure capable of being alive.

Chromosome

Threadlike DNA structures found in the nucleus of cells. Humans inherit one set from each parent.

Clone

A genetically identical copy; naturally occurs in some plants and bacteria.

Commensalism

A symbiotic relationship where one organism benefits and the other is neither harmed nor helped.

Confluencism

The interpretive lens of this anthology — a philosophy seeing spiritual meaning woven through scientific truth.

Crossing Over

The exchange of DNA between paired chromosomes during meiosis, creating genetic variation.

D

Death (Biological)

The end of cellular or organismal function, followed by decomposition and recycling into new life.

Decomposer

Organisms (fungi, bacteria, worms, vultures) that break down dead matter, forming the foundation of renewal.

Diffusion

Passive movement of molecules from high to low concentration.

DNA (Deoxyribonucleic Acid)

The genetic code of life — a four-letter alphabet (A, T, C, G) carrying inherited instructions.

Dormancy

A period in which seeds, bulbs, or organisms pause activity while potential remains alive.

E

Ecosystem

A community of organisms interacting with one another and with their environment.

Embryo (Seed Biology)

The miniature plant within a seed containing the blueprint of future growth.

Endosperm

Nutrient-rich tissue within a seed that feeds the embryo during germination.

Energy Pyramid

A diagram representing how energy decreases as it moves from producers to top consumers.

Enzyme

A biological catalyst that speeds up chemical reactions without being consumed.

Evolution (Microevolution)

Small-scale changes in populations across generations (e.g., phage–bacteria arms race).

F

Fertilization

The fusion of sperm and egg, restoring the full chromosome number.

Flow (Osmosis/Spirit)

Movement toward balance; water moving to where it is lacking.

Fungi

A kingdom of decomposers essential for nutrient cycling.

G

Gene

A segment of DNA coding for a trait.

Genetic Paradox

A trait harmful in one context but beneficial in another (e.g., sickle cell trait protecting against malaria).

Genome

All genetic material contained in an organism's cells.

H

Heat Loss (Energy Transfer)

Energy that radiates away in biological systems — not wasted but reshaping climate and sustaining life.

Homeostasis

Biological balance; the ability to regulate internal conditions.

I

Inheritance

Passing genes from parents to offspring.

Invagination (Cell Biology)

Folding inward of cell membranes during some developmental processes.

Integrated Witness Panel

A LTA/Living Biology structural element combining canonical, cultural, scientific, and modern insights.

L

Lab Report

A structured method of scientific inquiry using hypothesis, variables, and data.

Level of Organization

Biology's hierarchy: atom → molecule → cell → tissue → organ → organism → population → ecosystem → biosphere.

Lichen

A pioneer species in primary succession that breaks rock into soil.

M

Meiosis

Specialized cell division producing eggs and sperm, halving chromosome numbers.

Membrane

A semi-permeable boundary surrounding cells, regulating what enters and exits.

Mitochondria

Organelles producing cellular energy (ATP).

Mutualism

A relationship in which both partners benefit.

N

Natural Selection

The process where advantageous traits become more common in populations.

Nested Design

The concentric structure of life (cells within tissues, tissues within organs, etc.), symbolized in Chapter 3.

Nutrient Cycling

Movement of elements through decomposition, growth, and renewal.

O

Osmosis

Movement of water across membranes toward areas of lower water concentration.

Organism

An individual form of life: plant, animal, fungus, bacterium.

Organ System

Groups of organs working together (e.g., circulatory system).

P

Parasitism

A relationship where one organism benefits at the expense of another.

Phage Plaque

Clear zones on a bacterial culture where phages have infected bacteria.

Population

A group of individuals of the same species living in a single area.

Primary Succession

Ecosystem formation on new surfaces (lava rock, glacier retreat).

Protein

A chain of amino acids performing structural and catalytic roles.

R

r/K Strategies

Two survival strategies: **r-strategists** — high reproduction, low parental investment. **K-strategists** — few offspring, high care.

Reflection

A chapter element offering a spiritual interpretation of scientific truths.

Renewal

The continual cycle of life arising from decay, succession, and recycling.

S

Sacred Margin

A reflective prompt inviting the reader to practice the chapter's theme.

Sacred Scientific Method

A Confluencist framing of inquiry using observation, testing, reflection, and humility.

Seed Coat

The protective outer layer of a seed.

Secondary Succession

Recovery and regrowth in a disturbed ecosystem where soil remains.

Stem Cell

An unspecialized cell capable of becoming multiple cell types — a symbol of potential.

Surface Area–to–Volume Ratio (SA:V)

A measurement explaining why cells remain small; smaller cells exchange materials more efficiently.

Succession

The ecological process by which life reclaims disturbed or barren environments.

Symbiosis

Any close biological relationship: parasitic, mutualistic, or commensal.

T

Tissue

Groups of cells performing a shared function (muscle, nerve, epithelial).

Transcription / Translation

Processes by which DNA is expressed into proteins.

Treasure (Surface or Buried)

A metaphor for resources offered quickly (surface) or formed slowly under pressure (buried).

V

Volume (Cell Biology)

The internal space of a cell; grows faster than surface area, limiting cell size.

Virus

A non-cellular infectious agent that requires a host cell to replicate (including bacteriophages).

W

Witness (Chapter Section)

The scientific or natural-world example illustrating the chapter's theme.

Word in Flesh (DNA)

The metaphor describing DNA as sacred writing within every living organism.

REFERENCES & STREAMS OF WISDOM

This anthology weaves together:

Canonical Streams Bible passages woven throughout reflections.

Cultural & Ancestral Streams African proverbs, Indigenous teachings, Hebrew tradition, global wisdom.

Scientific Streams Biology, ecology, genetics, microbiology, systems thinking.

Confluencist Streams Reflections interpreting the sacred rhythms within science.

ABOUT THE AUTHOR

George Powell is an author, a retired Biology teacher, and a curator of human stories whose work centers on belonging, wonder, and the sacredness of life. Rooted in Confluencism — a philosophy of unity across streams of wisdom — his writing invites readers into a deeper conversation between science and spirit.

George is the founder of **George's Media LLC**, where he develops transformative works blending reflection, art, and education, including The Living Testament Anthology and this companion volume, The Living Biology Anthology.

ABOUT THE ILLUSTRATOR

Nick Jones is a visionary artist whose work blends luminous realism, spiritual symbolism, and scientific imagination.

His illustrations for Living Biology transform cells, ecosystems, and cosmic patterns into living parables — images that teach, comfort, and awaken wonder.

Nick's collaboration with George Powell continues a shared commitment to art that heals, guides, and reveals.

ABOUT THE ILLUSTRATIONS & DIAGRAMS

All illustrations and diagrams in this volume were created specifically for the Living Biology Anthology.

They reflect a unified visual language:

» Luminous realism

» Warm earth tones & gold light

» Emeralds, indigos, milk whites, dawn hues

» Science rendered as sacred story

Each chapter contains:

» A full-page illustration

» A symbolic diagram (where applicable)

» Small margin icons connecting themes across the book

LIVING BIOLOGY — VOLUME I

Designed and compiled by **George's Media LLC**

www.ingramcontent.com/pod-product-compliance
Lightning Source LLC
LaVergne TN
LVHW070911210926
848482LV00017B/1048